# METHODS OF WARFARE

## THE WAR ON TERROR

BY JESSICA GUNDERSON

CREATIVE EDUCATION

Published by Creative Education
P.O. Box 227, Mankato, Minnesota 56002
Creative Education is an imprint of The Creative Company
www.thecreativecompany.us

Design and production by The Design Lab
Art direction by Rita Marshall
Printed by Corporate Graphics in the United States of America

Photographs by AP Images (Denis Doyle, Kenneth Lambert), Corbis (David Howells, Reuters, JB Russel/Sygma, George Steinmetz), Defense Imagery (Cherie Chullen, Tech. Sgt. Larry Simmons/U.S. Air Force), Getty Images (AFP, Karen Ballard/Pool, Yuri Cortez/AFP, DOD, Ramzi Haidar/AFP, Mahmud Hams/AFP, Ed Jones/AFP, Scott Olson, Hrvoje Polan/AFP, Erik Simonsen, U.S. Navy), iStockphoto (Blackwaterimages, Joel Carillet, James Margolis, Andrew Robinson, Terraxplorer)



Library of Congress Cataloging-in-Publication Data
Gunderson, Jessica.
Methods of warfare / by Jessica Gunderson.
p. cm. — (The war on terror)
Includes bibliographical references and index.
Summary: An examination of the weapons, tactics, and strategies used in the ongoing war against Islamic extremists, from guerilla combat and suicide bombings to high-tech weapons and new laws.
ISBN 978-1-60818-101-8
1. Terrorism—Prevention—Juvenile literature. 2. War on Terrorism, 2001-2009—Juvenile literature. I. Title.

HV6431.G854 2011
355.02—dc22 2010033844

CPSIA: 110310 P01387

First Edition
9 8 7 6 5 4 3 2 1

# TABLE OF CONTENTS

In the years following the September 11 attacks, columns of light filled the space formerly occupied by the "Twin Towers" of New York's World Trade Center.

In the late 1980s, a conflict rooted in terrorism began to rear its head on a global scale. This strife pitted Islamic fundamentalists, radical religious **militants** springing primarily from nations in the Middle East, against the countries and culture of the Western world. Spilling across parts of four decades, this conflict—which came to be known from the Western perspective as "The War on Terror"—grew from bombings and guerrilla combat into the first large-scale war of the 21st century, marked by the infamous events of September 11, 2001, and intensive military campaigns in the countries of Afghanistan and Iraq.

Islamic **extremists** form groups such as the notorious al Qaeda with the purpose of driving out Western influence from the Middle East, and, to make their cause known, they perpetrate public acts of violence against Western citizens. Although such terrorist groups often lack resources and funding, their actions can cause widespread death and fear. To combat such terrorist acts, which can happen anywhere, Western governments such as the United States' have taken action both at home, with the development of new governmental departments and the passage of new security measures, as well as abroad, with the deployment of military forces armed with high-tech weapons. And as the War on Terror continues, both sides continue to adapt their methods of warfare.

# CHAPTER ONE

# GLOBAL JIHAD

Islamic extremists believe in the concept of jihad, which calls on Muslims, or believers in Islam, to combat enemies of their religion. Jihad is a duty that, according to one interpretation from Islam's sacred book the Qur'an (or Koran), can be fulfilled by the heart, the tongue, the hand, or the sword. The jihad of the sword involves waging war against the enemies of Islam, and Islamic extremists believe that those who die in combat against such enemies are **martyrs** who are guaranteed a place in paradise in the afterlife. Although many Muslims believe in non-militant jihad, Islamic extremists use the jihadist **ideology** in a violent effort to resist or remove Western influences through terrorist activities, with the goal of establishing an ideal Islamic **caliphate**.

Although many terrorist attacks against the West seem like senseless acts, almost all are committed with a singular aim in mind. Islamic terrorists have a common goal of driving Western and Jewish presence from Muslim lands. The state of Israel is a point of contention for many Islamic extremists. The Jewish people believe that Israel, including the city of Jerusalem, is their holy land; however, Muslims also view it as their holy land. After the **Holocaust** before and during World War II (1939–1945), which killed millions of Jews and drove others from their European homes, the **United Nations (UN)**

Jerusalem has long been a treasured and controversial site, as it is considered a holy city by people of the Islamic, Jewish, and Christian religions.

### KIDNAPPING

In the War on Terror, Islamic **insurgents** and terrorists often used kidnapping and hostage-holding to further their cause. In Iraq, Islamic militants, many connected to al Qaeda, kidnapped hundreds of Westerners, including American civilians and **contractors**. The victims were often killed in brutal ways; one prominent example of this was Nick Berg, a U.S. civilian working for the military, who was beheaded in a grisly murder that was filmed and broadcast. Like bombings and other attacks, such kidnappings and murders were meant to incite fear among Americans and their allies.

created Israel as a Jewish state, angering many Arab Muslims in the region. The land was split between Jewish Israel and Arab Palestine. Numerous terrorist incidents and all-out wars between Israel and Palestine have occurred in the half century since.

The U.S., an ally and supporter of **democratic** Israel, has been the target of numerous Islamic terrorist attacks over the last four decades, as have many other Western countries. Al Qaeda, one of the most prominent terror organizations, has perpetrated many of the more recent attacks. Because terrorist organizations such as al Qaeda are groups rather than governments or countries, their resources are often limited. They have no established military and, in some cases, not much money. Therefore, these terrorist groups rely on car and suicide bombs, roadside bombs, kidnapping, and **hijacking** as their primary means of warfare. Many of their attacks target civilians.

Al Qaeda, using its ideologies of jihad and martyrdom, has recruited many Muslims to its cause. Al Qaeda was formed in the late 1980s by a wealthy Saudi Arabian named Osama bin Laden.

After being driven from both Saudi Arabia and Sudan by those nations' governments, the group finally found refuge in 1996 in the barren, mountainous lands of Afghanistan, receiving the protection of the ruling Taliban government, which supported al Qaeda's philosophies. Bin Laden created camps in Afghanistan and around the world to train and preach to his followers. The training of al Qaeda recruits consists of prayer, physical exercise, and weapons practice. Recruits are taught how to operate such weapons as machine guns and rocket-propelled grenade (RPG) launchers—hand-held tubes that fire rockets or grenades designed to damage tanks or armored vehicles—and how to make bombs and handle poisons.

Osama bin Laden

Terrorist attacks are designed not only to cause death and destruction but also to draw widespread public and media attention. Most terrorist attacks occur in heavily populated areas, and the bomb is typically terrorists' weapon of choice, both because of its destructive potential and its visibility. Car bombs, which are rigged to a vehicle and set to detonate either via a remote trigger such as a cell phone or when the vehicle is started, are widely used by terrorists because they are relatively easy to manufacture; also, vehicles carrying the bombs are mobile and can transport a large amount of explosives without arousing suspicion. In 1993, an attack on the World Trade Center in New York City was carried out by a van loaded with explosives. Ramzi Yousef, a terrorist with ties to al Qaeda, drove the van into the parking garage beneath the World Trade Center, lit a 20-foot (6.1 m) fuse, and fled. The van exploded, killing 6 people and injuring more than 1,000. Yousef and his co-conspirators had hoped the explosion would collapse the two towers, but the structures remained stable.

Al Qaeda training camps have been attended not only by al Qaeda extremists but also by militants from organizations in Palestine and elsewhere.

## AL QAEDA TRAINING CAMPS

Al Qaeda has long used training camps to prepare its followers to perpetrate terrorist attacks. These camps are in Middle Eastern and Asian countries such as Pakistan, Afghanistan, and Iraq, as well as in other parts of the world, such as India, Sudan, and the Philippines. Inside many of the camps, recruits begin each morning with prayer. They then go through a series of physical exercises and learn how to construct bombs and operate guns and other weapons. An estimated 20,000 men trained at al Qaeda-operated camps between 1996 and 2001.

The suicide attack is another popular terrorist tactic, and because many recruits believe in martyrdom, they are more than willing to sacrifice themselves in such a way. The "9/11" terrorist attacks were the most devastating suicide attacks ever to occur on U.S. soil. On September 11, 2001, a crew of 19 al Qaeda terrorists hijacked 4 commercial airliners on the East Coast. After gaining control of the planes, the terrorists flew two of them into the World Trade Center towers, causing the skyscrapers to collapse and killing thousands of civilians and rescue workers. A third plane was flown into the **Pentagon**, killing nearly 200 more. The fourth plane, likely headed for the U.S. Capitol, crashed in a Pennsylvania field after passengers fought back against the hijackers.

The 9/11 attacks, called "Holy Tuesday" by the terrorists, were plotted by mastermind Khalid Sheikh Mohammed, a top al Qaeda member. Mohammed's plan was modeled after a 1994 hijacking of an Air France flight by the Armed Islamic Group of Algeria; in that case, the hijackers intended to fly the plane into the Eiffel Tower in Paris, but the French military stormed the plane during a fueling stop, foiling the plan. Mohammed proposed a similar attack on the U.S. to Osama bin Laden in 1996, and in 1998 or 1999, bin Laden approved the plan, providing financial support and recruiting participants. The 19 hijackers were chosen because of their grasp of the English language and their ability to obtain U.S. visas. The hijackers consisted of two groups: the pilots, who

The rubble of the collapsed World Trade Center towers was so massive that parts of it burned for three months while cleanup progressed.

## SECONDARY BOMB DEVICES

The use of secondary bomb devices is one tactic of terrorists that targets not only civilians but also the rescue workers who come to the victims' aid. Terrorists know that within minutes of an attack, there will be first responders—**paramedics**, firefighters, and law enforcement officers—on the scene. In the 1993 World Trade Center bombing, the terrorists planted a cyanide bomb, a deadly chemical, which was set to explode shortly after the first bomb, its target being rescue workers. Fortunately, the cyanide bomb was burned up in the first explosion.

would fly the planes, and the "musclemen," who would help overpower and control the flight crew and passengers. Three of the pilots, including the group's overall leader, Mohammed Atta, received flight training at an aviation school in Florida.

Bin Laden was aware that the 9/11 attacks would cause a global reaction; some experts believe that the eruption of war between the West and countries in the Middle East was his main intent. According to Bruce Riedel, who served in the U.S. **Central Intelligence Agency (CIA)** for nearly 30 years, al Qaeda's goal was to create a prolonged "bleeding war" that would deplete the U.S. and its allies of financial and military resources, thus driving Western presence from the Middle East and causing the isolation and collapse of Israel. For al Qaeda, Riedel stated, "the only way to undo the wrong perpetrated against the Arab world is not to negotiate a peace agreement or a bargain between the Israelis and the Palestinians, but to destroy Israel." Whether or not al Qaeda intended a full-scale war, the U.S. and its allies did offer a strong response to 9/11, quickly taking the fight to the terrorists with military might and advanced technology.

### QUTBISM

Qutbism (*kuh-TAH-bih-zum*) is an ideology based upon the writings of Sayyid Qutb, a prominent Egyptian thinker who wrote Islamic fundamentalist works in the 1950s and '60s. Quoted frequently by Islamic fundamentalists and jihadists, Qutb denounced the Western world and called for a true Islamic society, stating that Muslim nations and non-Muslim nations cannot coexist. Qutbism combines Islamic teachings of other thinkers such as Abul Ala Maududi and Hassan al Banna with Qutb's writings to justify the use of armed jihad to advance Islam.

Segments of the Islamic world—including some Palestinians—demonstrated their support for bin Laden and al Qaeda in the days following 9/11.

# CHAPTER TWO

# THE WORLD RESPONDS

The U.S. and other Western nations had been implementing strategies intended to prevent and defeat terrorism long before the 9/11 attacks. In 1977, after numerous terrorist incidents such as the failed New York City bomb plot of 1973 and the bombing of a flight en route from Israel to New York City in 1974, which killed 17 Americans, the U.S. launched a counterterrorist unit called Delta Force. Delta Force, modeled after the British Special Air Service, is an elite military group designed to combat terrorism using covert strategies and state-of-the-art weaponry and equipment, including high-powered sniper rifles and night-vision-equipped helicopters that fly without lights. Delta operators typically wear civilian clothing and are considered a unit separate from other military branches. Delta Force is so undercover that the U.S. government does not even officially acknowledge its existence—a confidentiality that is maintained in order to more effectively combat terrorism.

In addition to such military units, the U.S. created special counterterrorism units within the CIA beginning in 1986. The CIA, established in 1947, investigates and analyzes foreign **intelligence** that could affect national security and provides the president with a daily briefing about its discoveries. A network of operatives, or spies, planted around the globe helps gather intelligence. The CIA also works closely with foreign governments to root out terrorists and other national security threats. After 9/11, the CIA set up counterterrorist intelligence centers in more than two dozen countries, and CIA operatives worked together with foreign officials to capture or kill suspected terrorists.

An organization similar to the CIA, the Federal Bureau of Investigation (FBI) also aids in the deterrence of terrorism, but it deals primarily with domestic, rather than foreign, affairs. The FBI is made up of highly trained and specialized agents who often work alongside state and local police forces to protect U.S. citizens. In 1982, due to the rise in terrorist incidents around the world, the FBI made counterterrorism its fourth national priority; after 9/11, counterterrorism became its top priority. In New York City in 1980, the New York Police Department and the FBI created a Joint Terrorism Task Force (JTTF), and soon other local police departments around the country followed suit. A JTTF is made up of local investigators, FBI agents, and other federal agents, and its mission is to respond quickly to terrorist threats and incidents and to investigate and prevent future attacks.

### 1973 NEW YORK BOMB PLOT

Organized by Black September, a militant Palestinian terrorist group, the New York City bomb plot of 1973 was a plan to detonate three car bombs in the city during a visit by Israeli prime minister Golda Meir, with the intention of killing her and publicizing the Palestinians' grievances against Israel. The bombs failed to explode due to a circuitry problem, and the plot was discovered by U.S. agents. The FBI launched an investigation, which led to the arrest of plotter Khalid Al-Jawary.

The U.S. sent aircraft carriers such as the USS *Theodore Roosevelt* toward Afghanistan in September 2001 as it prepared for an attack on al Qaeda.

The terrorist attacks of 9/11 prompted sweeping investigative, military, and legislative action in the U.S. Within hours of the attacks, the FBI was able to identify the hijackers and link them to the terrorist network al Qaeda. U.S. president George W. Bush promised retaliation against the terrorists and the governments that harbored them. He warned the Taliban government in Afghanistan that if it did not hand over Osama bin Laden, the country could face military invasion. The Taliban refused, and Bush, with the support of the U.S. Congress, launched Operation Enduring Freedom and sent CIA paramilitary forces and the U.S. Army to the region. Great Britain, America's closest ally, also authorized a military campaign in Afghanistan.

Initial air strikes in October 2001 targeted Taliban military sites and terrorist training camps; the goal was to oust the Taliban from power and destroy al Qaeda hideouts, capturing or killing as many terrorists as possible. Ground troops soon joined in the fight, along with Afghanis from the Northern Alliance,

President George W. Bush

## THE TALIBAN

The Taliban, a Sunni Islamic political group formed in the late 1980s, gained control of Afghanistan's government in 1996. Adhering to a strict interpretation of religious law, the Taliban enforced rigid rules upon Afghani citizens and banned dancing, music, and television. Women were prohibited from education or employment, and men were required to wear long beards. Disobedience of these rules was punishable by beatings, imprisonment, and even death. The Taliban approved of al Qaeda's philosophies and gave the group a safe haven in Afghanistan until U.S.-led forces overthrew the Taliban government.

The mountainous landscape of Afghanistan, featuring abundant cave systems, made finding and striking al Qaeda and Taliban targets difficult.

a collection of militant groups that opposed the Taliban, and forces from the nations of the **North Atlantic Treaty Organization (NATO)** collectively called the International Security Assistance Force (ISAF). Within two months, the Taliban strongholds of Mazar-i-Sharif, Kabul, and Kandahar fell, and many Taliban leaders fled or were captured.

The U.S. considered the fall of the Taliban a victory; however, al Qaeda terrorists were more difficult to root out. Groups of terrorists, called cells, had spread out, and many terrorists took cover in the mountains, hiding in caves and underground bunkers or tunnels. The U.S. military responded by deploying "bunker-buster" bombs that, when dropped, could penetrate deep into the earth before exploding, thus destroying even deeply buried hideouts. In mid-November 2001, ground forces entered Tora Bora, a mountainous cave complex near the eastern border with Pakistan where bin Laden was believed to be hiding. The Battle of Tora Bora was a major step in eradicating many terrorists, as more than 200 were captured or killed in the fierce fighting that ensued. However, bin Laden continued to evade capture or death.

As the next several months passed, the ISAF and U.S. forces—armed with the technological advantages of their bunker-busting bombs and "smart" bombs that struck targets with great precision using a **global positioning system (GPS)**—seemed headed toward victory over al Qaeda and Taliban militants, who could not match the firepower, training, or technology of the Western forces. By early 2002, about 10,000 U.S. troops were stationed in Afghanistan, whereas an estimated 500 al Qaeda and Taliban fighters remained, the rest having fled or been killed or captured. In February 2002, the new national Afghan army began training with American troops as the U.S. prepared to hand over control of the country to the newly democratized Afghanis. Although bin Laden had not been captured, the al Qaeda network had been weakened and scattered.

Saddam Hussein

The War on Terror was not over, however. In fact, it soon expanded in scope as suspicions arose against Saddam Hussein, the president of Iraq. President Bush and his

## B-2 SPIRIT STEALTH BOMBER

The B-2 Spirit is an enormous U.S. bomber aircraft that uses stealth technology to covertly drop bombs. Weighing in at 160,000 pounds (72,575 kg), with a wingspan of 72 feet (22 m), the B-2 relies on its sleek design to avoid detection by radar. The bomber is made of radar-absorbing material, and its thin, triangular shape, along with gray coloring that blends into the sky, allows it to fly undetected in enemy territory. The B-2 was used in bombing runs in the opening stages of the Afghanistan and Iraq wars.

## WEAPONS OF MASS DESTRUCTION

Weapons of mass destruction (WMD) are weapons capable of causing widespread human casualties or destruction. Nuclear weapons, which use the enormous energy obtained from the splitting of **nuclei** for their explosive power, are one kind of WMD. Another type is the biological weapon, which spreads viruses, bacteria, or toxins. Chemical weapons are similar to biological weapons except that they disperse deadly chemical toxins. Nuclear weapons are delivered via missiles or dropped bombs, whereas chemical or biological WMD can be distributed more discretely via mailed packages, spray, or silent gas bombs.

The shock-and-awe invasion (a strategy more formally known as "rapid dominance") of Iraq in March 2003 left Baghdad shrouded in smoke.

administration believed that Iraq was developing weapons of mass destruction, with possible intentions to sell the weapons to terrorist groups such as al Qaeda. After UN inspectors failed to uncover the alleged weapons in December 2002, Hussein was given until March 17, 2003, to disclose their location or face invasion. When Hussein struck a defiant tone and did not turn over the weapons, the U.S.—with the support of forces from Britain, Australia, Denmark, Poland, and Spain—attacked Iraq. The allied forces pummeled targets in the capital city of Baghdad with a quick, destructive barrage of more than 500 cruise missiles—a so-called "shock-and-awe" assault that was intended to frighten and demoralize the Iraqi forces, leaving them unable or unwilling to fight. Although Hussein's army was soon defeated, chaos, insurgency, and warfare would continue for years, spreading U.S. resources thin as terrorist cells gained strength in Iraq and regrouped in Afghanistan.

# CHAPTER THREE

# THE SPREAD OF TERRORISM

Although the terrorists and insurgents were outmatched in terms of manpower and technology, they proved to be a formidable enemy, due both to their willingness to die and their knowledge of the region and terrain. When American and ISAF forces invaded Afghanistan in 2001, the mountains surrounding Tora Bora were an al Qaeda stronghold. Together with Taliban fighters, al Qaeda militants hid in the plentiful caves, underground bunkers, and tunnels. In December 2001, CIA operatives in the area confiscated an al Qaeda radio that was broadcasting bin Laden's speeches to his followers and pointed to his being in Tora Bora.

The terrain was difficult for the Americans and their allies to navigate, but because al Qaeda and Taliban forces were so familiar with the mountains, it was easy for them to slip away from their attackers and to take shelter underground during heavy air strikes. Between December 4 and 7, the U.S. dropped 700,000 pounds (317,515 kg) of explosives on the mountains. One bomb, the massive BLU-82, nicknamed "Daisy Cutter," weighed 15,000 pounds (6,800 kg) alone and destroyed an area as large as 300 yards (274 m) in radius. The al Qaeda militants fought back with their own artillery of RPGs and planted bombs. During the Battle of Tora Bora, 220 al Qaeda members were killed and

Anti-Taliban Afghani militants look on from afar as U.S. aircraft blast the mountains of Tora Bora with bombing runs in December 2001.

### SUNNIS AND SHIITES

In the Islam religion, two major **sects** exist: Sunni and Shia. In Iraq, during Saddam Hussein's rule, Sunni leaders comprised much of the government, despite the fact that more than 60 percent of the country was Shiite. After the fall of Hussein, Iraq held elections that placed many Shiites in leadership roles, which angered the Sunni people. Bombings and violence became rampant between the two sects. Although the U.S.-led coalition supported both sects, some Sunni groups rebelled against the coalition forces as well as the new government.

about 50 more were captured; however, bin Laden escaped, allegedly crossing the border into Pakistan.

After being driven from Tora Bora, al Qaeda and Taliban fighters regrouped in another rugged region of Afghanistan, the Shahi-Kot mountains. During the U.S.'s Operation Anaconda in March 2002, the militants used guerrilla hit-and-run tactics, firing at U.S. forces with machine guns, RPGs, and small shell launchers called mortars, and then retreating into caves or underground tunnels. Many militants were killed, but, as in Tora Bora, many others slipped away into Pakistan.

The porous Afghanistan-Pakistan border would continue to vex allied forces. Beginning in late 2001, many members of al Qaeda and the Taliban crept across the border into Pakistan, finding haven with sympathetic Pakistani civilians, then re-crossed the border into Afghanistan to fire on U.S. and ISAF troops. Because the Pakistani government had declared **sovereignty**, it would not allow U.S. or ISAF forces into its country or its airspace, giving al Qaeda members a measure of safety as they traveled back and forth across the international border.

When U.S.-led forces invaded Iraq in 2003, the instability and upheaval that ensued made the country a prime location for terrorist groups such as al Qaeda. Although Iraq's government did not have proven ties to al Qaeda, in the aftermath of the 2003 invasion, al Qaeda terrorists found a haven among some sympathetic citizens, such as insurgent Sunni Muslims. Al Qaeda in Iraq was led by Abu Musab al-Zarqawi until he was killed by a U.S. air strike in 2006. After his death, the group steadily declined in influence due to multiple factors, including a U.S. troop surge in 2007 and a loss of support from Sunni Iraqis; however, in late 2009, the group appeared to regain some strength.

After the Afghanistan invasion, al Qaeda decentralized its organization. Rather than being headquartered primarily in Afghanistan, small but powerful al Qaeda cells spread throughout the Middle East and around the globe, with cells in places such as Somalia, Sudan, and the Philippines. This strategy of decentralization made it more difficult for al Qaeda

Abu Musab al-Zarqawi

## TERRORISTS AND INSURGENTS

In the War on Terror, U.S.-led forces have battled both insurgents and terrorists, and although the terms are sometimes used synonymously, they have inherent differences. Terrorists attack civilians to call attention to their cause or philosophy, whereas insurgents fight against a specific government or **occupation** in their own country. The insurgents in Iraq and Afghanistan employed many of the same guerrilla tactics as the terrorists, and many terrorists joined the insurgency. In Afghanistan, the Taliban was a large component of the insurgency, and in Iraq, Sunni militants comprised the largest insurgent movement.

Emergency personnel in Madrid, Spain, organize the bodies of people killed when bombs exploded in four commuter trains on March 11, 2004.

members to be found and captured, and the spread of al Qaeda philosophies around the world strengthened the organization. Many men trained at al Qaeda training camps, and although some who went on to carry out attacks may not have been official al Qaeda members, they could often be linked back to the training camps. A 2004 bomb attack in Madrid, Spain, which killed 191 people, and a 2005 bombing in London, England, which killed 56, were carried out by Islamic extremists who had questionable ties to al Qaeda, although the group stated it was proud of the attacks. Al Qaeda often accepted responsibility for such attacks, eager for the publicity and notoriety.

In addition to publicly claiming responsibility for terrorist attacks, al Qaeda and

### CHEMICAL WARFARE

One of the greatest fears in the Western world is that terrorists will use chemical and biological weapons to cause widespread casualties and spread new waves of terror. In September 2001, shortly after 9/11, letters containing the deadly powder anthrax were mailed to U.S. media and government offices, killing 5 people and infecting at least 17 others. Although the anthrax attacks were later attributed to an American scientist rather than to Islamic terrorists, the episode increased worries that terrorists could adopt such methods of attack.

other terrorist groups utilized the Internet and other forms of technology to make their name and cause known. In 2010, there were more than 200 English-language Web sites urging Western Muslims to take up jihad against the U.S. and its allies. Nearly 500 Web sites in Arabic languages also advocated global jihad. These sites often contained sermons and writings from Muslim **clerics**. By offering such information in widely known languages such as English, radical Islamic groups hoped their message would reach and recruit Muslims in all corners of the globe and not just those in the Middle East. Paul Eedle, a British journalist and expert on the Middle East, called the Internet an ideal tool for al Qaeda. "Al Qaeda has much wider ambitions than just setting off explosives," he said. "It is trying to mobilize the whole Muslim world against the West."

The Internet also allowed prominent leaders such as bin Laden to broadcast messages to their followers and media outlets around the world. Prior to the rise of the Internet in the mid-1990s, bin Laden often sent his videotapes to Al Jazeera, an independent Middle Eastern news network, but advancements in technology

Specialists wearing hazardous materials suits stand guard near the U.S. Capitol in Washington, D.C., during the anthrax scare of 2001.

### THE HURT LOCKER

The 2008 film *The Hurt Locker* illustrates the dangers faced by a U.S. Army Explosive Ordnance Disposal (EOD) unit, or bomb squad, during the early stages of the Iraq War. Based upon real-life observations made by war journalist Mark Baol, the fictional film follows team leader William James, garbed in a thick bombsuit, as he skillfully diffuses IEDs under intense, stressful situations. Winner of the 2010 Academy Award for Best Picture, *The Hurt Locker* was the first highly acclaimed film about the Iraq War.

A U.S. soldier looks at the damage done to a police checkpoint in Kandahar City, Afghanistan, by an IED left in a wheelbarrow.

made it easier for him and other terrorists to upload videos to Web sites, thus making the message instantly and globally available.

As the War on Terror played out through the 2000s, insurgents in both Iraq and Afghanistan, who often aligned themselves with terrorist groups such as al Qaeda, increasingly used improvised explosive devices (IEDs) to attack the U.S.-led **coalition** of military forces. IEDs (also called "roadside bombs," because they are frequently planted along roadways) are homemade bombs that are often buried or placed inconspicuously, such as under vehicles or in garbage cans, and can be set to detonate by remote control or with a timer. Through 2010, IEDs were the number-one killer of troops in Afghanistan and Iraq.

# CHAPTER FOUR

# A NEW WAY FORWARD

In Iraq in 2004, insurgents controlled the city of Fallujah, located about 40 miles (64 km) west of Baghdad, and coalition forces made the removal of insurgents from the area a major objective. Two key battles for Fallujah were fought, the first in April 2004, in which the coalition failed to gain control of the city. In the second, beginning in late October, coalition forces—supported by the Iraqi military—surrounded the city to ensure that no insurgents were able to escape. Air and artillery strikes began on November 8. A few days later, Marines moved through neighborhoods of the city, waging dangerous, house-to-house urban combat to capture or kill the insurgents and confiscate their weapons. The battle ended in a coalition victory, with more than 1,200 insurgents killed and thousands of weapons seized, with the coalition suffering 38 fatalities. However, insurgent presence remained strong in the region, and in 2006, rampant insurgent attacks led to another military sweep of the city.

From invasion onward, the foremost priorities of the troops in Iraq and Afghanistan were to stomp out terrorists and insurgents and to protect civilians as the countries moved toward stable democratic governments. One of the most important goals was to win the "hearts and minds" of Iraqi and Afghan civilians by providing protection, medical services,

U.S. Marines ready themselves for close combat as they search a house in Fallujah as part of the second battle for control of the city in 2004.

## BLACKWATER

In 2003, the U.S. government hired Blackwater USA (renamed Xe Services in 2009), a private security company, to provide services in Iraq. Many Blackwater contractors were former military or police force employees who acted as security guards for **diplomats**, construction workers, and government officials. Some contractors also helped train the Iraqi police. Blackwater came under close scrutiny after a 2007 incident in which guards allegedly shot and killed 17 Iraqi civilians without provocation, although the guards maintained that they had acted in self-defense. The controversy caused Blackwater to be replaced in Iraq by a similar company named Triple Canopy.

A Blackwater helicopter flies over Baghdad.

and food and water supplies while minimizing "collateral damage"—a military term for the accidental harm done to civilians or their property during combat. In the words of U.S. General Stanley McChrystal, "If the people are against us, we cannot be successful. If the people view us as occupiers and the enemy, we cannot be successful."

The deadly guerrilla tactics employed by insurgents, such as the use of IEDs, compelled U.S.-led forces to use every technological advantage available to protect themselves. Because insurgents were likely to attack at any time, not just during full-fledged battle, ground troops often wore body armor, consisting of a vest and shoulder panels made of a ceramic material that could absorb impact from gunfire or shrapnel. Because of the high temperatures common in the desert environments of Iraq and Afghanistan, the troops' gear was continually modified and redesigned to be as lightweight and cool as possible. Troops patrolled in armored high-mobility multipurpose vehicles (Humvees) made of strong steel and glass that could withstand gunfire. Humvees were typically equipped with large-caliber machine guns, night-vision scopes, radios, and smoke or flare launchers.

Another valuable technological advantage of the U.S. military was the unmanned aerial vehicle (UAV), an aircraft that could fly remotely without human pilots. UAVs such as the Scan Eagle were equipped with cameras, sensors, and monitoring systems that collected surveillance videos and data from the area to be transmitted back to a military base. Troops then used this information to strategize an attack before ever entering the area. Some UAVs, called Predator drones, could also drop bombs or fire missiles on targets and, because the aircraft held no human crews, did not incur loss of life if shot down by the enemy.

Yet despite such high-tech advantages, it became apparent as early as mid-2003 that a victory by the U.S.-led coalition in Iraq was not imminent. The Iraq War had thrust the nation into chaos, with fighting erupting among various sects, most notably the Sunni insurgency. Al Qaeda, too, had moved into the area and found refuge among Sunni insurgents. By the end of 2006, as the number of coalition troops killed in Iraq topped 3,000, people around the world were growing increasingly dissatisfied

Predator drones were created more for surveillance than attack, but beginning in 2001, they were increasingly outfitted with Hellfire missiles.

## SPECIAL AIR SERVICE

Great Britain's Special Air Service (SAS), an elite military unit formed in 1941 during World War II, responds to terrorist threats at home and abroad and has been used as the model for similar units in other countries such as Australia and the U.S. Each SAS squadron is composed of 64 men; within each squadron, troops of 16 specialize in different warfare locales—air, water, mountain, and ground. Britain deployed SAS squadrons to Iraq in 2003 as part of a task force to help battle al Qaeda.

### CLEAR AND SECURE

In outlining the 2007 troop surge strategy for Iraq, U.S. president George W. Bush noted that a primary goal in Baghdad would be to "clear and secure neighborhoods." First used in the Vietnam War in the 1960s, this strategy calls for troops to clear areas of enemy combatants and then remain in the area to maintain peace. Prior to the surge, there had not been enough troops to secure the areas in fiercely contested cities such as Baghdad and Fallujah; the troops would clear a particular zone and then move on, which allowed insurgents to creep back into the area.

with the way the Iraq War was being handled. According to 2006 assessments by American intelligence agencies, the chaos that emerged in the country after the U.S.-led invasion helped provide a new breeding ground for terrorists. Terrorism analyst Peter Bergen remarked that the war in Iraq had "energized al Qaeda, its affiliated groups, and like-minded jihadists around the world. What has happened in Iraq is what bin Laden could not have hoped for in his wildest dreams."

Whether or not the invasion of Iraq had helped or hindered terrorism, the fact remained that coalition troops were entrenched in the country, and the challenge now facing the involved Western nations was how to achieve victory there. In January 2007, President Bush announced a strategy he called "The New Way Forward," which deployed 20,000 new troops to Iraq and lengthened the tour of 4,000 Marines. Many of the troops were sent to Baghdad, where they were embedded into Iraqi military units to help maintain security in the city.

The troop surge did appear to quell violence in Baghdad as well as in the western Anbar province, where insurgency against the coalition and the Iraqi government was a major problem. In September 2008, the U.S. handed over control of the region to Iraqi forces, and in June 2009, it gave responsibility for Baghdad to Iraqi troops, although thousands of U.S. troops remained in the country. Coalition countries had steadily been withdrawing troops since 2004, and Britain was the last to remove its troops in June 2009, leaving the Americans as the sole foreign presence in Iraq.

In Afghanistan, despite initial victories in 2001 that had removed the Taliban from power and captured many al Qaeda members, instability within the new Afghani government and lack of a clear strategy for U.S. and ISAF troops eventually led to a resurgence of the Taliban and al Qaeda. From 2005 onward, al Qaeda and Taliban militants increased their attacks on U.S. and ISAF forces in Afghanistan. After being elected in 2008, new U.S. president Barack Obama proposed a revised strategy in the War on Terror—one that relied more upon diplomacy and intelligence and less upon broad military force. Following up on

President Barack Obama

a campaign promise, he ordered the withdrawal of combat troops from Iraq in August 2010 and concentrated American forces in Afghanistan instead.

Shortly after taking office, Obama authorized sending additional troops to Afghanistan, primarily along the country's border with Pakistan, the stronghold of terrorists and Taliban militants. In December 2009, he ordered 30,000 additional troops to be deployed to Afghanistan, with the promise that after 18 months, the troops would begin to come home. Obama also called for worldwide support. As of 2011, more than 40 countries had troops stationed in Afghanistan as ISAF forces under NATO's control, and Obama asked these countries to keep their troops in the region. Although bin Laden was finally located in Pakistan and killed in May 2011, the conflict went on.

When the U.S. was attacked on 9/11, President Bush proclaimed that the War on Terror would be a war unlike any other and could continue for years to come. His words proved correct as the conflict widened in scope, encompassing the nations of Afghanistan and Iraq, and as terrorist cells evaded defeat and continued to proliferate. As battles were fought and years passed, one truth became very clear—even in an age of instant global communication and ever-evolving technology in surveillance and weapons, terrorism is an enemy and concept not easily defeated.

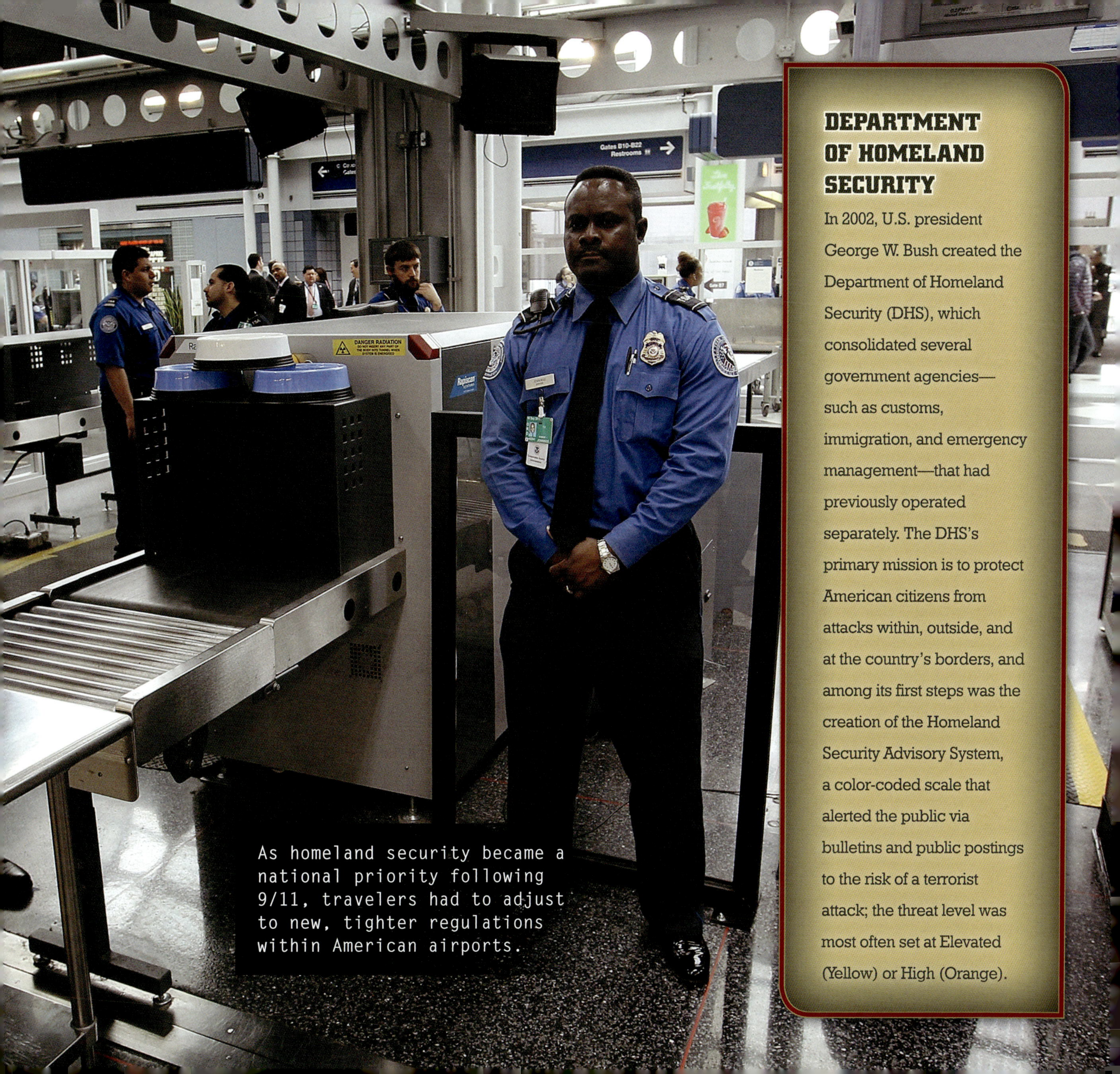


As homeland security became a national priority following 9/11, travelers had to adjust to new, tighter regulations within American airports.

## DEPARTMENT OF HOMELAND SECURITY

In 2002, U.S. president George W. Bush created the Department of Homeland Security (DHS), which consolidated several government agencies—such as customs, immigration, and emergency management—that had previously operated separately. The DHS's primary mission is to protect American citizens from attacks within, outside, and at the country's borders, and among its first steps was the creation of the Homeland Security Advisory System, a color-coded scale that alerted the public via bulletins and public postings to the risk of a terrorist attack; the threat level was most often set at Elevated (Yellow) or High (Orange).

# ENDNOTES

**caliphate** — an Islamic form of government led by a head of state called a caliph; a goal of many Islamic extremists is to establish a caliphate throughout the world's Islamic-majority countries

**Central Intelligence Agency (CIA)** — a U.S. agency established in 1947 that uses intelligence—gathered by a network of operatives, or spies—to prevent and fight threats to the nation's security

**clerics** — religious leaders or priests; in the context of the War on Terror, the term usually refers to Islamic religious leaders

**coalition** — an alliance of individuals or groups who join together for a common cause

**contractors** — people who provide specific materials or specialized services, such as security or construction, for a predetermined fee

**democratic** — describing a form of government in which the leaders are elected by the people; the term also describes social equality of the individuals in society

**diplomats** — people who participate in nonhostile negotiations or interactions carried out between countries working toward alliances, treaties, or agreements

**extremists** — people who hold views that are radical or far from the traditional or common view

**global positioning system (GPS)** — a high-tech system that provides precise location and time information based upon data generated by satellites

**hijacking** — forcing the violent takeover of a vehicle, especially a plane, bus, or other mode of public transportation

**Holocaust** — the systematic slaughter of six million European Jews (along with other minorities) by German Nazis between 1933 and 1945

**ideology** — the system of ideas, theories, and philosophies of a group of people

**insurgents** — people who fight or otherwise actively participate in a revolt or uprising against a government or ruling force

**intelligence** — information concerning political or military matters, including potential acts by an enemy or possible enemy

**martyrs** — people who sacrifice something or die for their religion or beliefs; in the world of Islam, the word "shahid" is used for Muslims who die while fulfilling religious commandments or waging war in the name of Islam

**militants** — people who use aggression or combat in support of a cause

**North Atlantic Treaty Organization (NATO)** — an alliance of 28 countries in North America and Europe that provides political or military support to protect its member countries

**nuclei** — central portions of atoms (the most basic units of matter) that contain a positive charge; when split or fused, energy is released

**occupation** — the holding and control of an area by a foreign military or force

**paramedics** — medical workers who perform emergency care on sick or injured patients before and during transport to a hospital

**Pentagon** — a huge, five-sided building near Washington, D.C., that is the headquarters of the U.S. Department of Defense

**sects** — groups within a larger group that adhere to distinctive doctrines or beliefs

**sovereignty** — a state of self-government and independence from foreign control

**United Nations (UN)** — an organization with representatives from 192 nations that deals with international law, human rights, and economic progress, and aims to maintain peace between nations through communication

# SELECTED BIBLIOGRAPHY

Honigsberg, Peter Jan. *Our Nation Unhinged: The Human Consequences of the War on Terror*. Berkeley, Calif.: University of California Press, 2009.

Kepel, Gilles, and Jean-Pierre Milleli, eds. Translated by Pascale Ghazaleh. *Al Qaeda in Its Own Words*. Cambridge, Mass.: Belknap Press of Harvard University Press, 2008.

Riedel, Bruce. *The Search for al Qaeda: Its Leadership, Ideology, and Future.* Washington, D.C.: Brookings Institution Press, 2008.

Rubin, Barry, and Judith Colp Rubin. *Chronologies of Modern Terrorism*. Armonk, N.Y.: M. E. Sharpe, 2008.

Sciutto, Jim. *Against Us: The New Face of America's Enemies in the Muslim World*. New York: Harmony Books, 2008.

Tyler, Patrick. *A World of Trouble: The White House and the Middle East—From the Cold War to the War on Terror*. New York: Farrar, Straus and Giroux, 2009.

U.S. Department of Defense. "U.S. Department of Defense Official News Website." http://www.defenselink.mil/.

The White House Archives of President George W. Bush. "Fact Sheet: The New Way Forward in Iraq." White House News and Policies. http://georgewbush-whitehouse.archives.gov/news/releases/2007/01/20070110-3.html.